YOUR KNOWLEDGE HAS VALUE

- We will publish your bachelor's and master's thesis, essays and papers

- Your own eBook and book - sold worldwide in all relevant shops

- Earn money with each sale

Upload your text at www.GRIN.com and publish for free

Fahim Rahman

Food Consumption Pattern and Dietary Diversity of the FSNSP Covered Households

GRIN Publishing

Imprint:

Copyright © 2010 GRIN Verlag GmbH
Print and binding: Books on Demand GmbH, Norderstedt Germany
ISBN: 978-3-656-34543-5

This book at GRIN:

http://www.grin.com/en/e-book/202765/food-consumption-pattern-and-dietary-diversity-of-the-fsnsp-covered-households

Food Consumption Pattern and Dietary Diversity of the FSNSP Covered Households

6/10/2010

Acronyms:

EC   :   European Commission

FSNSP   :   Food Security and Nutrition Surveillance Project

JPG   :   James P Grant Public Health School

HKI   :   Helen Keller International

HIES   :   Household Income and Expenditure Survey

MoF & DM   :   Ministry of family welfare and Disaster Management

RED   :   Research and Evaluation Division

WFP   :   World Food Programme

Content:

<u>**List of Tables**</u>

<u>**List of Titles**</u>

Chapter-1: Introduction

The extent of hunger and food insecurity in a country is an imperative welfare indicator (Anand and Harris 1990). The Food and Agriculture organization of the United Nations estimates around 800 million people worldwide to be food insecure and they are not retained in the boundary of the developing world. Measuring food insecurity at the individual/household level rather than national level differs from the more traditional approach of identifying food insecurity as the inadequacy of aggregated supply of and accessibility to food (Busch and Lacy 1984). Recently, however dissatisfaction with many of the available measures of food security has led to the use of direct measure of food insecurity (Maxwell 1995; Maxwell et al. 1999; Wolfe and Frongillo 2001) such as household food consumption data (based on recall). Household food consumption has been defined as the total amount of food available for consumption in the household, generally excluding the food taken outside unless prepared at home (Klaver, Knuiman et al. 1982). It serves as an indicator of food security as well as a distal proxy indicator for poverty. Information about food consumption and dietary diversity is important from the programmatic point of view as it has the potentiality to be used as effectively to detect change, modify, or improve programs activities (Jalal et al. 2009).

Today, food insecurity, and under-nutrition persist as the most serious and chronic public health problems facing Bangladesh. Emphasizing in the key areas of food security and nutrition, the government of Bangladesh has felt the need for a comprehensive, nationwide surveillance system. In response to this crucial need for accurate and timely surveillance data on nutrition and food security in Bangladesh, the European Commission (EC) funded the National Food Security Nutritional Surveillance Project (FSNSP). Food security and Nutrition Surveillance Project (FSNSP) is a collaborative follow up project of both Helen Keller International (HKI) and James P Grant Public Health School (JPG) of BRAC University. The central aim of the project is to measure household's food diversity and food security in Bangladesh. Through this project data on socio-economic status, water, sanitation and hygiene, food availability and consumption pattern, morbidity pattern and immunization of children, information regarding reproduction and mother health were also collected.

Research and Evaluation Division (RED) was advanced as a multi-disciplinary sovereign research unit within the framework of BRAC. BRAC RED has also initiated a study with comparatively small sample size while the project of FSNSP was going on. The aim of the study conducted by BRAC RED was to test the validity of the food security and food diversity questionnaire used by FSNSP. This paper provides information about the consumption pattern and dietary diversity of the studied households using food frequency questionnaire. This paper also provides information of poverty among the studied households using per capita per day calorie consumption as a proxy indicator and the percentage of households with food insecurity.

Chapter-2: Methodology

1) Study Design:

The study, carried out by BRAC RED was cross sectional in design. It was a descriptive assessment of the dietary consumption pattern in Bangladesh. For the study we used a quantitative method to collect dietary information in a short period of time to fulfill the objectives of the study.

2) Site selection:

The study households were selected among the households that were selected by FSNSP project in the first round. FSNSP targeted to collect data from 61 districts out of 64 districts under 7 divisions of Bangladesh in two rounds (31 districts in first round and the remaining 30 districts in the second round). FSNSP excluded 3 districts of Chittagong hill tracts due to communication problem. For our study we had also excluded the Chittagong hill tracts and randomly select 18 districts out of the 31 districts in the first round of the FSNSP study.

3) Sample size of the study:

In the first round of the FSNSP project, a total of Seven hundred and twenty households were randomly selected from 18 districts in Bangladesh. Twenty households from each village were selected for the interview where FSNSP team selected 30 households from each village. Among the study samples 9 households were from those who provided incomplete and missing values hence these 9 samples were excluded from the analysis. Besides these, in baseline analysis 5 households had calorie consumption above 5000 kcal/capita/day were considered outliers (Wiesmann et al. 2009) and excluded from the subsequent analysis. So the sample size of the study was 706.

4) Data collection process:

A structured questionnaire, based on the three-day recall method was applied to gather dietary information. Data was collected from the female members of the households, who are usually more informed about food purchases, intra-household food allocation, and cooking. The respondents were asked to recall all food items that they had consumed within the last three days prior to the interview. A checklist of food items was used by the enumerators to help the respondents recall the names and amount of the food consumed. The checklist also helped them calculate the number of household members who had eaten during those days. The quantity of food consumed at the household level was estimated by the standard measures cups that were provided to the enumerators. They measured the raw weight of the foods in grams.

5) Data analysis:

Household 24 hour dietary recall method was used to obtain the total amount of food available for consumption in the household. An adult individual present at every meal within the one day period was counted as a full-time consumer, with the value of one. The number of family members and visitors eating each meal was recorded and the "man value" was assigned to each person, weighted according to the age and sex of the household members. Rome scale was used in the case for assigning man value, where males >14y were assigned a value of 1.0; females >11y and boys 11-14y, a value of 0.90; children 7-10y a value of 0.75; children 4-6y, a value of 0.40; and children <4y, a value of 0.15 (Moller Jensen et al. 1984). The amount of each food consumed by the entire household was divided by the corresponding total man value to provide food intakes per person. This approach produces a better estimate of the adequacy of the household food intake (Gibson 2007).

The food items consumed were pooled into 11 groups (Cereals, potato, vegetables, pulses, dairy, edible oil, meat & eggs, fish, Spices, fruits, sugar & molasses) for detailed information of the consumption pattern of different food items.

6) Data management:

All completed questionnaires were checked for inconsistency and errors by the supervisor before sending to the BRAC/RED head office for computerization. A researcher prepared a coding manual and data entry layout and thereafter data were entered using the SPSS 11.5 software. The same software was used for data cleaning and analysis. To convert the amount of consumed food per person into nutrients, nutrient composition of Bangladeshi food tables was used (Darton-Hill et al. 1988). Calculating software was developed to convert food intake data into nutrients.

7) Recruitment and Training

Dietary enumerators were trained on techniques of the intra-family food distribution survey including weighing raw and cooked food using standard measuring tools. They received training in the use of food models, verbal probes, and the interview procedure. Before collecting data 12 enumerators were given extensive 5-day intensive training on survey techniques, use of the questionnaire of the 24-hour recall and actual weighing methods and calculation of food intake at the individual level. To measure house hold food amount a set of standard measuring cups and spoon were provided to them. In each day of the data collection period each enumerator collected data form 4 households.

8) Time line:

	Jan 10'	Jan 10'	Feb 10'	Mar10'	Apr10'	May10'	Jun10'	Jul 10'	Aug10'	Sep10'
	1-15	16-31	1-20	1-30	1-10	11-20	21-31	1-10		
proposal writing	■									
Data collection		■	■							
Data entry and coding				■						
Data cleaning & analysis					■	■	■	■		
Report writing									■	

Chapter-3: Results

The per capita per day intake of food from different food groups within the households is shown in table-1. Around 91% of the food consumed per capita per day was from plant-based food whereas rest was from animal-based foods.

Among all the food groups per capita daily intake of cereals was highest followed by the consumption of vegetables and potato. Cereals, mainly rice, contributed to almost 585g by weight of average food consumed daily (1104g). Among the vegetables, non-leafy vegetables (186g) were consumed more than leafy vegetables (15g). Pulses consumption was the lowest among all the food groups (6g).

Within the animal food sources consumed (97g), fish was the most frequently consumed animal food and contributed to almost 59g by weight whereas beef and chicken contributed to 3g and 8g respectively. Edible oil contributed to approximately 18g of the daily food consumed by weight per day.

Table-1: Per capita daily food intake of the households

Food groups (g)		Mean±SD (N=706)	Percentage (N=706)
Cereals		584.6±158.3	53.0
	Rice	513.3±165.5	46.0
	Atta	31.4±14.8	3.0
	Others	39.9±18.1	3.0
Potato		116.7±64.4	11.0
Vegetables		201.3±127.2	18.0
	Leafy vegetables	15.2±8.9	1.0
	Non leafy vegetables	186.1±109.4	17.0
Pulses		6.4±3.1	
Dairy*		23.9±17.1	2.0
Edible oils		18.4±11.3	1.0
Meat, eggs*		13.9±9.0	1.0
	Beef	2.7±1.3	
	Chicken	8.4±6.1	
	Eggs	2.8±1.6	
Fish*		58.8±34.5	5.0
Fruits		51.2±34.1	5.0
Sugar/Molasses		7.2±5.7	
	Sugar	5.4±4.2	
	Molasses	1.8±1.5	
Spices		22.1±12.4	2.0
	Onion	19.8±11.3	
	Others	2.3±1.1	
Total		1104.5±321.8	
Plant-based Food		1007.9±478.1	91.0
Animal-based Food		96.6±60.6	9.0

*Animal-based food

Intake of energy, macro-nutrients, iron, and calcium are summarized in table-2. Average calorie intake was estimated at 2675 kcal per day. Carbohydrates contributed to approximately 80% of the total calorie consumption whereas protein and fat contributed to 10% each. Mean calcium intake was only around 450mg whereas iron intake was about 34mg.

Table-2: Mean per capita daily intake of the nutrients of the households

Nutrients	Mean±SD (N=706)
Energy (kcal)	2675.0±786.8
Carbohydrate (g)	543.5±142.8
Protein (g)	68.3±19.5
Fat (g)	29.2±19.1
Calcium (mg)	449.2±273.4
Iron (mg)	33.6±10.4

Cereals (mainly rice), vegetables (both leafy and non-leafy), potato, and fish constituted the main food items consumed within households (Table-3). Edible oil was consumed by almost 98% of the households. Only 3% households consumed beef. Egg consumption was higher than beef and chicken. Among the spices group onion was consumed by almost 99% households. Slightly higher percentage of households got protein from plant sources (24%) than combinedly from meat, poultry, and egg groups (20%).

Table-3: Percentage of households using particular food groups

Food groups	Percentages (N=706)	
Cereals	100	
Potato	86.5	
Vegetables	99.0	
Leafy vegetables		98.6
Non leafy vegetables		97.8
Pulses	23.5	
Dairy	30.4	
Edible oils	97.7	
Meat, eggs	20.2	
Beef		2.6
Chicken		5.2
Eggs		12.4
Fish	64.1	
Fruits	47.1	
Sugar/Molasses	38.8	
Spices	99.0	
Onion	98.5	
Others	14.0	

The households' daily energy intake was mostly from plant-based foods (Title-1) which accounted for as high as 96% for energy. Cereals contributed to 78% of per capita daily energy intake followed by edible oil (6%), potato (5%), and vegetables (4%). Cereals were the highest contributor for protein also. Among the animal source of protein fish contributed the highest. Meat, poultry, and eggs contributed to 5% of total protein intake per day per capita which is even low from the protein from vegetables.

Title-1: Contribution of food groups to energy and Protein consumption daily per capita

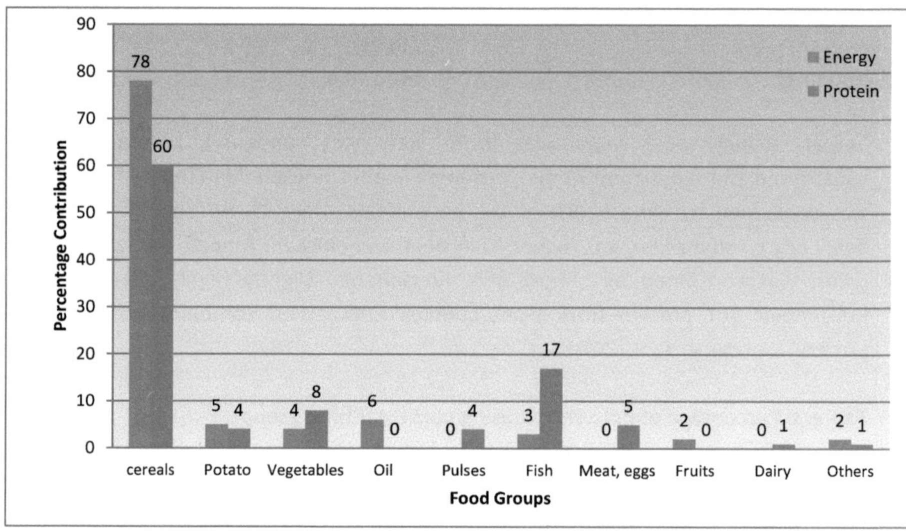

Among the households cereals were the leading source of iron (Title-2) which accounted 80% of the total iron consumed from food. After cereals vegetables were the second most shared source of iron but accounted only 12%. Almost 50% of the calcium intake per day was contributed by fish followed by vegetables (19%), cereals (15%), and Dairy (8%).

Title-2: Contribution of food groups to iron and calcium consumption daily per capita

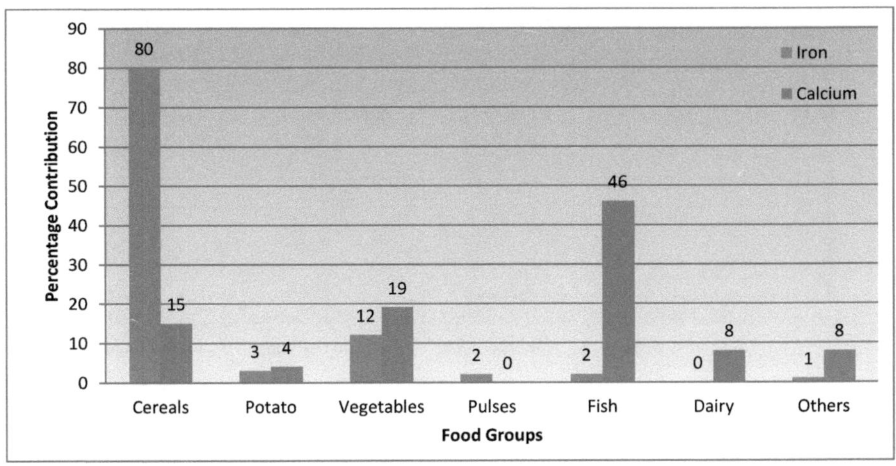

One of the ways used to define poverty is through identifying the average calorie Intake per capita per day. Approximately 7% of the households consumed below 1805kcal and hence were claimed as extremely poor and 18% of households consumed below 2122kcal/day and were classified as poor (Table-4).

Table-4: Percentage of households within different energy intake level

Energy intake level	Percentages (N=706)
<1805 kcal/day	7.2
1805-2122 kcal/day	10.7
>2122 kcal/day	82.1

Among all the households only 17% (N=706) consumed food daily from all the 6 food groups (Title-3). On the other hand 12% of the households consumed food from only 3 food groups out of the 6 groups and these included cereals, vegetables, and oils. Rest of the households consumed food from cereals, vegetables, oils, and protein (mainly fish) group.

Title-3: Proportion of households achieving dietary diversity (six food groups)

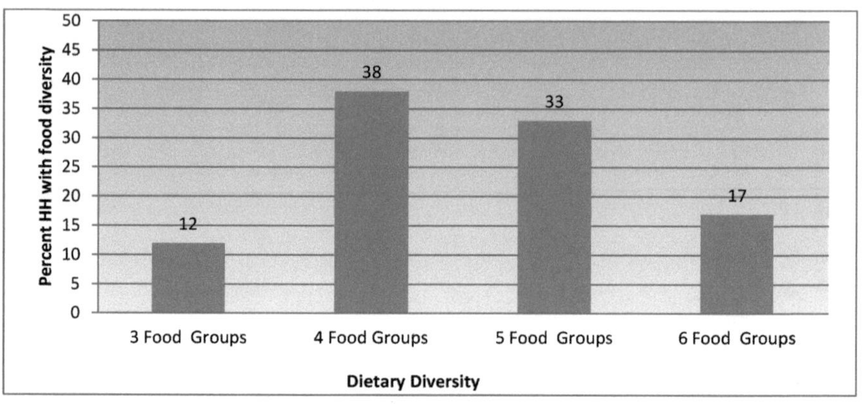

To estimate the percentage of people who were food insecure a variable of three groups of different calorie consumption was constructed using the calorie consumption cutoffs shown in Table-5 (WFP 2005). Only 2% household were food insecure and around 16% household were in the borderline of food insecurity.

Table-5: Percentage of households within thresholds calorie consumption group

Calorie consumption kcal/capita/day	Profile	Percentage of HH (N=706)
< 1470	poor	2.0
1470 - <2100	Borderline	15.6
>= 2100	Acceptable	82.4

Chapter-4: Comparison with National Data

We compared the amount of food consumed by the households with the national consumption as reported by Household Income and Expenditure Survey (HIES) by Bangladesh Bureau of Statistics (BBS 2007) in table-5. The intake of cereal was higher by 115g which was 25% higher from the cereal intake in 2005 HIES survey. Vegetables, edible oil, and fish intake was also higher from the 2005 HIES survey. Consumption of pulse and combined meat and eggs was lesser by 7.8g which is lees by 55% and 6.9g which was less by 33% respectively from the 2005 HIES survey.

Table-5: Comparison of per capita mean amount of food intake of the survey households with Household Income and Expenditure Survey.

Food groups	Study survey	2005 HIES National	2000 HIES National
Cereals	584.6	469.2	486.7
Potato	63.3	55.0	116.7
Vegetables	157.0	140.5	201.3
Pulse	6.4	14.2	15.6
Dairy	23.9	32.4	29.7
Edible oil	18.4	16.5	12.8
Meat, eggs	13.9	20.8	18.5
Fish	58.8	42.1	38.5
Fruits	51.2	32.5	28.4
Sugar/Molasses	7.2	8.1	6.9
Total	1104.5	947.7	893.1

The amounts of major food groups consumed by the households have been compared to the recommended intake (MoF & DM, BD) for a Bangladeshi individual per capita per day (Title-4). Cereals intake was much higher and vegetables intake was slightly higher compared to the recommended intake. Fruits, edible oil, and animal protein intake were just about half of the recommended intake. Pulse intake was much lower (about 90%) than the recommended intake per capita per day.

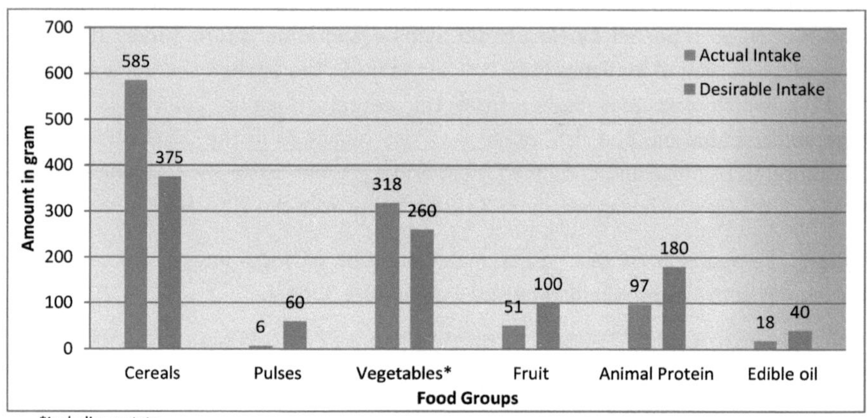

*Including potato

Chapter-5: Discussion

The major objective of the study conducted by RED, BRAC was to test the validity of the food security and food diversity questionnaire used by FSNSP. This paper mainly discuss about the food consumption pattern and dietary diversity of the studied households by using household food record method (Gibson, 2005). Besides these, this paper also describes the poverty level of the studied households using per capita per day calorie consumption as a proxy indicator (BBS, 2003). The percentages of households with food insecurity and in the borderline of food insecurity were also measured using threshold level of per capita per day calorie consumption (WFP, 2007). This threshold level for defining food insecurity was also used by others (Weisden, 2008). For dietary data intake of the studied households, 24-hour dietary recall method was used which was validated by other studies also (Haseen, 2006; Maxwell et al., 1999; Wolfe and Frongillo, 2001; Willett, 1998).

The food consumption pattern of the studied household was analyzed in details. The study showed that consumption of cereals were highest among all the food consumed followed by potato, non-leafy vegetables, fish and fruits. Beef, egg, pulses, and leafy vegetables were consumed in the lowest amount per day per capita. Around 91% of the consumed food per day per capita was from plant-based food. Rice and other non-cereals plants are the major sources of energy for the people of Bangladesh. Cereals alone account for almost 80% of the energy requirements of average Bangladeshi diet (Murshid, 2008) which was also similar in our study.

Average daily per capita calorie intake was estimated at 2675 kcal. Carbohydrates contributed to approximately 80% of the total calorie consumption whereas protein and fat contributed to 10% each. This figure depicts that the household's energy intake from carbohydrate was much higher where they were deficient in protein and fat intake. The higher amount of per capita per day cereal intake was found in other studies also (Jalal et al. 2009; Haseen, 2006).

According to the HIES survey conducted by BBS (2007) the required per capita per day amount of cereals is 375g whereas our study showed that the studied household's per capita per day cereals intake was 585g, which was higher by 210g. The higher intake of cereals than the recommended intake level per day per capita was also found by Jalal et al. This higher amount of cereals intake as well as plant based food intake increases the probability of micronutrients deficiency by reducing micronutrient absorption (west et al. 2002; Gibson et al. 1998). To achieve complete volumes of food these studied households adding cheaper cereals based food in their diet and thus the quality of their diet is compromised.

Animal- based foods were not the major source of protein of the households. Around 60% of the protein intake per day per capita derived from cereals which are low quality protein. Only one fourth of the total protein consumed per day was derived from animal based foods. Fish contributed to around 74% of the animal protein. Fish, especially small fish, is the most important animal food in the Bangladeshi diet, supplying animal protein and micronutrients with high bioavailability (Roos, 2001; Haseen, 2006).

Though per capita daily iron intake of the studied households (33.6mg) was higher than the recommended intake per day per capita, 80% of the consumed iron derived from the cereals are non-heme iron with very low bio-available because of the presence of inhibitory factors (Brune, 1992). Per capita per day calcium intake was only 37% of the recommended intake of 1200mg per day. Haseen (2006) in her study also found similar results in the consumption pattern of calcium. Fish alone provide 50% of the calcium intake of the studied households per day.

Cereals (mainly rice), vegetables (both leafy and non-leafy), potato, and fish constituted the main food items consumed within households. Only 17% of the households took food from 6 diverse food groups each day. 12% consumed food from only 3 groups which were cereals vegetables and edible oil; these households do not consume any food from animal source. Food insecurity was measured based on the calorie consumption cutoffs (WFP, 2007). Only two percent was found food insecure and 16 % households were in the borderline.

Though 82% of the households were food secured based on the calorie consumption cut-offs, still these households were deficient in their micro-nutrient intake. These household's members did not consume food from all the six food groups that is maximum households food were not from diversified.

The argument of food security has not given enough attention to the quality and safety of food and the need for a balanced diet. Ensuring ample supply of energy dense cereal based food can only address the problem of hunger but others manifestations of nutritional shortage cannot be solved by energy dense foods alone but requires adequate intake of micro-nutrients.

References:

Anand S, and C. J. Harris (1990). 'Food and Standard of Living: An Analysis Based on Sri Lankan Data', in J. P. Drèze and A. K. Sen (eds), *The Political Economy of Hunger*, vol. 1. Oxford: Clarendon Press, 297–350.

BBS (2003). Report of the household income and expenditure survey 2000. Bangladesh Bureau of Statistics, Planning division, Ministry of Planning, Government of Bangladesh.

BBS (2007). Report on the household income and expenditure survey 2005. Bangladesh Bureau of Statistics, Planning division, Ministry of Planning, Government of Bangladesh.

Brune, M., Rossander-Hulthe'n, L., Hallerberg, L., Gleerup, A. & Sandberg, A. (1992) Iron absorption from bread in humans: inhibiting effects of cereal, fibre, phytate and inositol phosphates with different numbers of phosphate groups. Journal of Nutrition, 122, 442-449.

Busch, L., andW. Lacy (1984). 'What Does Food Security Mean?' in L. Busch andW. Lacy (eds), *Food Security in the United States*, 1–10. West View Press, Boulder, Colorado.

Darton-Hill I, Hassan N, Karim R, Duthie MR (1988). Tables of Nutrient Composition of Bangladesh Foods. Helen Keller International, Bangladesh.

Gibson RS (2005). Principles of Nutritional Assessment. New York: Oxford University Press.

Gibson RS, and Ferguson EL (1998). Assessment of dietary zinc in a population. Am. J. Clin. Nutr. 68 (suppl): 430S–434S.

Haseen F (2006). Change in Food and Nutrient Consumption Among the Ultra Poor: Is the CFPR/TUP Programme Making a Difference? CFPR/TUP Working Paper Series No. 11: BRAC and Aga Khan Foundation Canada.

Jalal CSB, Choudhury N, and Suliman M (2009). Food Consumption Pattern and Dietary Diversity. Pathways Out of Extreme Poverty: Findings from round I survey of CFPR phase II. BRAC RED: 199-210.

Klaver W, J Knuiman *et al.* (1982). Proposed definitions for use in the methodology of consumption studies. The diet factor in epidemiological research. J Hautvast and W Klaver. Wageningen Ponsen and Loogen: 77-85. (Euronut report 1)

Maxwell D. (1995). 'Alternative Food Security Strategy: A Household Analysis of Urban Agriculture in Kampala'. *World Development*, 23: 1669–81.

Maxwell D, Ahiadeke C, Levin C, Armar-Klemesu M, Zakariah S, and Lamptey G, (1999). 'Alternative Food-Security Indicators: Revisiting the Frequency and Severity of 'Coping Strategies'. *Food Policy*, 24: 411–29.

Moller Jensen O, Wahrendrof J, Rosenqvist A. Geser A (1984). The reliability of questionnaire-derived historical dietary information and temporal stability of food habits in individuals. American journal of Epidemiology 120: 281-290.

Murshid KAS, Khan NI, Shahabuddin Q, Yunus M, Akhter S, Chowdhury OH (2008). Determination of Food Availability and Consumption Patterns and Settings up of Nutritional Standard in Bangladesh. Bangladesh Institute of Development Studies.

Roos N (2001). Fish consumption and aquacultutre in rural Bangladesh. Department of human nutrition. Denmark: The Royal Veterinary and Agriculture University, Frederiksber. (Doctoral thesis)

West CW, Eilander A, van Lieshout M (2002). Consequences of revised estimates of carotenoid bioefficacy for dietary control of vitamin A deficiency in developing countries. Journal of Nutrition.132 (9 Suppl): 2920S–6S.

Wiesmann D, Bassett L, Benson T, Hoddinott J. Validation of the World Food Programme's Food Consumption Score and Alternative Indicators of Household Food Security. IFPRI Discussion Paper 00870: June 2009.

Wolfe W, and Frongillo E. (2001). 'Building Household Food-Security Measurement Tools from the Ground Up'. *Food and Nutrition Bulletin*, 22: 5–12.

World Food Programme (WFP), (2005). Emergency food security assessment handbook. 1st ed. Rome.